WATER IN RIVERS AND LAKES

Isaac Nadeau

The Rosen Publishing Group's
PowerKids Press™

New York

To Joe Brodsky

Published in 2003 by The Rosen Publishing Group, Inc.
29 East 21st Street, New York, NY 10010

First Edition

Editor: Gillian Houghton
Book Design: Maria E. Melendez

Photo Credits: Cover, title page, pp. 8 (right), 16, 20, and all page borders © Digital Vision; pp. 4, 7, 8 (inset), 11, 12, 15, 19, illustrations by Maria E. Melendez.

Nadeau, Isaac.
Water in rivers and lakes / Isaac Nadeau.
    p. cm. — (The Water cycle)
Includes bibliographical references (p.   ).
ISBN 0-8239-6266-0 (lib. bdg.)
1. Rivers—Juvenile literature. 2. Lakes—Juvenile literature. 3. Watersheds—Juvenile literature. 4. Hydrologic cycle—Juvenile literature. [1. Rivers. 2. Lakes. 3. Watersheds. 4. Hydrologic cycle.] I. Title.
GB1203.8 .N34 2003
 551.48'2—dc21

                                2001006171

Manufactured in the United States of America

CONTENTS

ocean

river

lake

Less than 1 percent of all the water on Earth is found in streams, rivers, ponds, and lakes. However, these bodies of water are important sources of freshwater. All living things on Earth need water to process food and waste and to breathe.

THE MANY FACES OF WATER

Water is one of the most amazing **substances** on Earth. It takes on several forms and moves in many different ways. Unlike any other natural substance on Earth, water is commonly found in three states, or forms. It occurs as a solid, a liquid, and a gas. The same **molecule** of water that hangs frozen in an icicle might someday float in a cloud as **water vapor**, run along the banks of a river, or seep deep under ground.

This movement of water on Earth and in the sky is called the water cycle. Many living things, including people, make their homes near supplies of freshwater, such as rivers and lakes. Though lakes and rivers make up only a tiny part of the total amount of water on Earth, they are an important part of life to the people and the other living things that depend on them.

WE ALL LIVE IN A WATERSHED

Wherever you live on Earth, whether you live in a desert, in a forest, in a city, or on an island in the middle of the ocean, you live in a watershed. A watershed is an area of land that receives rain, snow, or hail and carries it downhill. In some watersheds, the water collects in a lake or a **wetland**, where it might **evaporate** or soak into the ground. Most watersheds are part of even larger watersheds. The water in the Mississippi River, for example, comes from thousands of smaller rivers and streams. Together these rivers and streams form a watershed that covers 1.2 million square miles (3.1 million sq km) of land. Each river and stream forms its own smaller watershed. Water that flows from one watershed to another is called runoff. As the water from the small watersheds continues to flow downhill, runoff collects in larger rivers.

Rocky Mountains

Appalachian Mountains

Mississippi
River

The pull of gravity carries water to an area of lower ground between two ridges. This area of land makes up a watershed (inset). On a larger scale, the Mississippi River watershed drains all of the water between the Rocky Mountains and the Appalachian Mountains (above). Hundreds of streams and rivers empty into the Mississippi River.

A river's source often is referred to as its headwaters (right) and can be thousands of miles from the river's mouth, or end. Many smaller rivers and streams, called tributaries, come together to make a large river, such as the Mississippi River (inset).

HOW RIVERS BEGIN

If you were to follow a river as far upstream as you could, you would eventually reach its source, or beginning. Some rivers are fed by melting glaciers. Most rivers have their source where **groundwater** comes to the surface. In some cases, this happens at a **spring**, where water bubbles out through a crack in the ground. The ground around a spring is so **saturated** that water comes to the surface without a pump or a well.

Water flows downhill because of the pull of gravity. As the water flows, it washes away the ground beneath it to form a channel. Through time the channel is cut deeper, forming a stream or a river. A stream is a small river. Most of the world's great rivers start out as tiny streams. A small stream or river that flows into a larger one is called a **tributary**. Thousands of tributaries may flow into a long river.

HOW RIVERS CARVE THE LAND

Wherever water flows, it shapes the land beneath and around it. Water in a stream or a river carries bits of earth, called sediment, with it. Sediment comes in all sizes, from grains of sand to huge boulders. As these bits of earth scrape against the bottom of a river channel, they cause more sediment to fall into the river. The process of carving the banks of a river deeper and wider is called **erosion**. Soft, sandy earth is more easily eroded than hard rock. Rivers that flow over soft earth usually carry more sediment and move slower than do rivers that flow over hard rock. The larger and faster a river is, the larger the sediment it can carry. Through millions of years of erosion, the Colorado River has carved a valley in Arizona known as the Grand Canyon. It is 1 mile (1.5 km) deep and 18 miles (29 km) wide.

As running water carves out a new channel, sediment is carried away (above left). If there is a regular supply of water, the channel will become a river. Rocks and sediment carried by the water (inset) will erode the riverbed, making it deeper and wider (above right). The power of the rushing water also erodes the banks.

The headwaters, also known as the upper course of a river, often contain rapids (above left). The water flows quickly over boulders and large sediment. The main body, or middle course, of a river flows slower (above right). Here the river is wider and deeper than it is near the headwaters. At the mouth, or lower course, of a river, the water empties into the ocean. The river is widest and deepest here (right).

MAKING A DEPOSIT

*E*ventually all sediment carried by a stream or a river is deposited, or set down. Sediment is deposited in places where a river slows down and no longer has the strength to carry it farther. This can happen when the river bends sharply, becomes wider and shallower, or empties into a lake or an ocean. Heavier sediment is deposited first, often creating rapids. Smaller pieces can be carried even in slow-moving water. Deposited sediment creates many kinds of landforms. Sandbars and gravel bars are small islands of either sand or gravel deposited in the middle of a wide, slow-moving channel. Deltas are wide areas of deposited sediment that form at the mouth, or the end, of a river. At a delta, the river breaks into many small channels, which cut their way through many years of deposited sediment.

FLOODS

Many people build their homes close to the banks of rivers. Some people like to be close to the river because it is beautiful. Other people make their living by using the river's natural resources. In many cases, the soil close to the river is **fertile**, making it a good place to grow crops. Seasonal floods contribute to these conditions. A flood occurs when a river overflows its banks. Each time a river floods, sediment is deposited outside its banks. Often this sediment is rich in nutrients. Over time many layers of sediment are deposited, creating a wide plain of rich soil called a floodplain. Many of the world's most productive farming communities are on floodplains. Unfortunately, floods also can be very dangerous to people. When floods are especially high, water can fill the streets and can destroy homes.

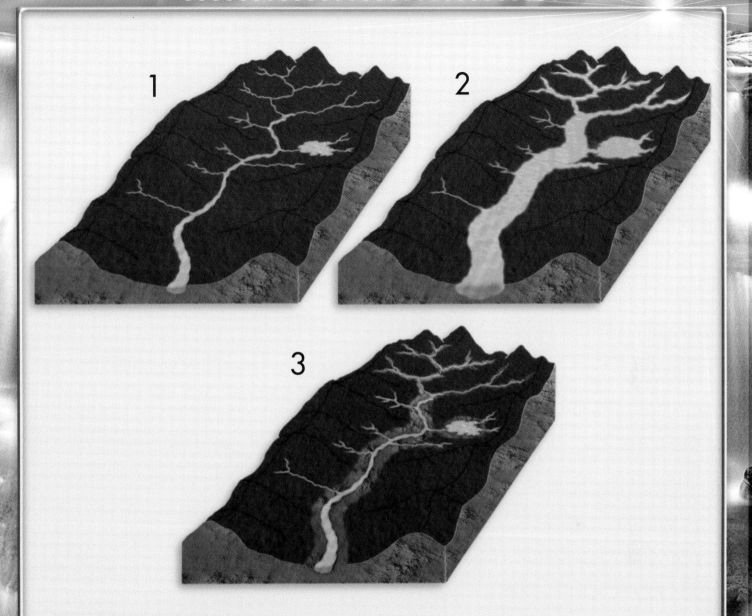

During dry seasons a river runs within its banks (1). Floods occur for a variety of reasons. Often seasonal rains increase the amount of water in a watershed. Sometimes a river becomes choked with sediment and is forced to overrun its banks. The water level in the river rises and overflows the river's banks (2). When the floodwaters retreat, they leave behind a layer of nutrient-rich sediment (3), making these floodplains ideal for farming.

Lakes range in size from small mountain lakes (above) to vast inland seas. The world's largest lake is central Asia's salty Caspian Sea, which covers 143,244 square miles (371,000 sq km).

WHAT IS A LAKE?

A lake is a large body of water collected in a **depression** in Earth's surface. Most lakes contain freshwater, although some lakes, such as the Great Salt Lake in Utah, contain salt water. Lakes are fed by streams, rivers, **precipitation**, and groundwater. Many lakes are formed by the collection of water at the end of a river. These lakes often are the sources for rivers that flow downhill from them. Lakes occur all around the world. They can be high in the mountains or deep in desert valleys. Some lakes are small enough to throw a stone across. Some are so large you can't see across them, even on a clear day. Some lakes are only a few feet (m) deep, and others can be as much as 3,000 feet (914 m) deep.

THE LIFE CYCLE OF A LAKE

*L*akes are formed in a variety of ways. **Crater** lakes are formed in the craters made by volcanoes. Some lakes are formed in cracks in Earth's crust caused by earthquakes. Some lakes are formed behind dams, where people have blocked a river to harness its power for electricity. Lakes do not last forever. A lake might dry up or drain away. It might become filled with sediment and bits of dead plants and animals. This process is called **eutrophication**. As natural waste fills the lake, it becomes shallower. Eventually the lake can become a body of water called a marsh. A marsh is so shallow that plants can reach from the muddy bottom to the water's surface. As the plants grow in size and number, the marsh becomes shallower and drier. In time a forest might grow in an area that once was a lake.

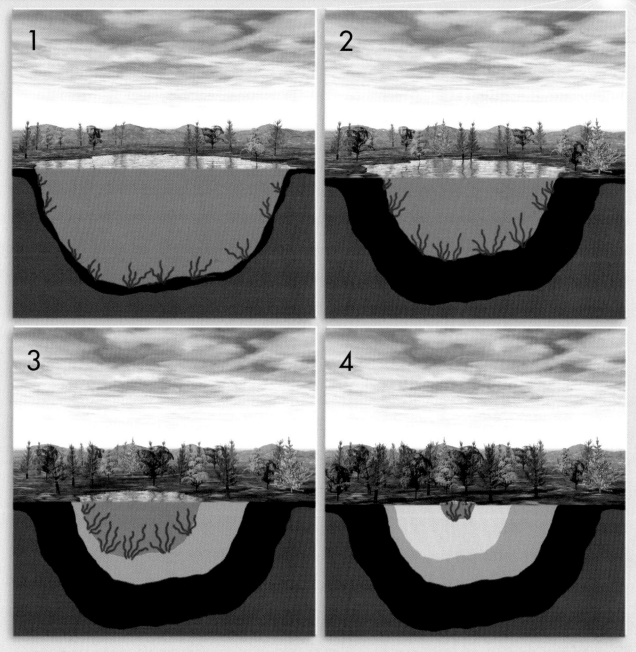

The floor of a young lake (1) is covered by a thin layer of mud and is dotted with water-dwelling plants. As time passes, sand, soil, plants, and rotting dead animals form a thick, muddy bottom (2). The lake becomes a pond (3), then a marsh, and, finally, a fertile lake plain (4).

Each year Americans take more than 1.8 billion trips to our nation's beaches, rivers, and lakes, such as this lake located at the foot of Mount Washington in Oregon. However, this popularity takes its toll. The Environmental Protection Agency has found that 39 percent of all U.S. lakes are significantly polluted.

RIVERS, LAKES, AND PEOPLE

People depend on water from rivers and lakes in many ways. Rivers are used to transport goods across great distances. Some rivers are dammed to create electricity. Water from lakes and rivers is used to **irrigate** crops. People use rivers and lakes for recreation, such as fishing and canoeing. Some rivers and lakes are even believed to be sacred, or holy. The Ganges River in India is a river that is sacred to **Hindus**. Millions of people travel to the Ganges every year to bathe in its waters. The **Incas** believed that Lake Titicaca in South America was the sacred birthplace of their culture. People have special relationships with rivers because of the many benefits rivers bring and because of the beauty and the mystery that people find in them. The water in lakes and rivers is a small part of the world's water, but it is a very important part of people's lives.

21

TAKING CARE OF YOUR WATERSHED

Rivers and lakes are important parts of watersheds. Rivers and streams carry water and sediment from one place to another. Lakes, rivers, and streams provide special habitats, or places to live, for many different kinds of plants and animals, including humans. We can have serious effects on our watersheds. Today people all around the world are learning about the watersheds they live in and how they can take care of them. Farmers are finding ways to reduce the amount of **pollutants** and natural wastes that run off into lakes and streams. Many dams have been removed or rebuilt to let rivers flow freely again. We can learn new ways to enjoy the water we love while keeping it safe for the other creatures that depend on it.

GLOSSARY

crater (KRAY-ter) A hole in the ground, shaped like a bowl.

depression (dih-PREH-shun) An area of Earth's surface that is lower than the surrounding areas.

erosion (ih-ROH-zhun) The wearing away of something.

eutrophication (yoo-troh-fuh-KAY-shun) The filling in of a lake with sediment and the bodies of dead plants and animals.

evaporate (ih-VA-puh-rayt) To change from a liquid to a gas.

fertile (FUR-tuhl) Good for growing plants.

groundwater (GROWND-wah-tuhr) Water that is found underground, where all of the air spaces in the soil and rock are filled with water.

Hindus (HIN-dooz) People who believe in the Hindu religion of India.

Incas (ING-kuhz) A group of people from the western part of South America.

irrigate (EER-uh-gayt) To carry water from one place to another in order to water crops.

molecule (MAH-lih-kyool) A tiny building block that makes up a substance.

pollutants (puh-LOO-tants) Human-made wastes that harm the environment.

precipitation (prih-sih-pih-TAY-shun) Any form of moisture that falls from the sky.

saturated (SA-chuh-rayt-ed) Completely filled with something, usually a liquid.

spring (SPRING) An area where groundwater flows out at Earth's surface without a well.

substances (SUB-stan-sihz) Any materials that take up space.

tributary (TRIH-byoo-tehr-ee) A stream or river that flows into a larger stream or river.

water vapor (WAH-tuhr VAY-pur) The gas state of water.

wetland (WET-land) An area of shallow water where many plants and animals live.

INDEX

WEB SITES

Due to the changing nature of Internet links, PowerKids Press has developed an online list of Web sites related to the subject of this book. This site is updated regularly. Please use this link to access the list:

www.powerkidslinks.com/wc/rivlak/